CITY PARK

BY
WENDY DAVIS

Children's Press

A Division of Grolier Publishing
New York London Hong Kong Sydney
Danbury, Connecticut

Created and Developed by The Learning Source

Designed by Josh Simons

Illustrations by Arthur John L'Hommedieu
Photo Credits: David M. Dennis/Tom Stack & Associates: 12; Steve Elmore/Tom Stack: 1, 4-5; Travis Evans/Unicorn Stock Photos: 6; Sharon Gerig/Tom Stack: 15 (right); John Gerlach/ Tom Stack: 28; Russell R. Grundke/Unicorn: 23; Thomas Kitchin/Tom Stack: front cover, 26; Tom & Pat Leeson: 10, 21; MacDonald Photog./ Unicorn: 14; Tom McCarthy/Unicorn: 7; Tom & Deeann McCarthy/Unicorn: 29; Joe McDonald/ Tom Stack: 24; Rod Planck/Tom Stack: 9, 11 (left), 13 (inset); Marshall Prescott/Unicorn: back cover; Milton Rand/Tom Stack: 15 (left); Ted Rose/Unicorn: 3; Bob Rozinski/Wendy Shattil/ Tom Stack: 13, 27; Lynn M. Stone: 8, 32; SuperStock, Inc.: 18-20, 22, 25; Gary Vestal/Leeson: 11 (right).

Library of Congress Cataloging-in-Publication Data
Davis, Windy.
City park / by Wendy Davis.
P. cm. — (Habitats)
Summary: A city park is shown to be home to many different forms of animal life, from insects to birds and mammals.
ISBN 0-516-20741-5 (lib. bdg.) 0-516-20370-3 (pbk.)
1. Urban ecology (Biology)—Juvenile literature. 2. Urban animals—Juvenile literature. [1. Urban animals. 2. Urban ecology (Biology) 3. Ecology.] 1. Title. II. Series: Habitats (Children's Press)
QH541.5C6D35 1997
577.5'6—dc21 97-28741
CIP
AC

Printed in the United States of America
8 9 10 R 11 10 09 08 07

The buildings of the city stretch skyward, reaching up toward the clouds. Down below, crowds of people jam the sidewalks—talking, working, hurrying to where they need to go.

Here, even the air seems busy. Dark smoke rises in great swirls from cars and buses. Streets sing with the sounds of honking horns, chattering voices, and heavy machinery.

Everything in the city rushes and races and roars—everything except a bright patch of green. It is a park, a place where living things of all kinds can escape the noise and crowds.

In the park, trees cast their shade while birds and animals go about the business of living. Human visitors come, too. Sometimes they sit quietly and close their eyes for a short rest. Other times they look around to find a whole world hidden from the city beyond.

The park's plants and trees provide much more than bright colors and a place out of the sun. They also give the city clean, fresh air.

Trees and plants make their own food out of sunlight, carbon dioxide, and water. In the process, the trees and plants give off oxygen.

People and animals use oxygen in order to breathe. As a result of breathing, people and animals give off carbon dioxide. So what one group gives off, the other group uses to survive.

Beneath the leaves there is action of another kind. In the ground are decomposers such as worms, bacteria, and other organisms. They break down pieces of dead plant and animal matter, making them part of the earth. This helps create the rich soil that trees and plants need to grow strong.

Insects of all sizes are everywhere. Bumblebees and yellow jackets buzz about looking for sweet, sugary juices inside the flowers. Meanwhile, a bright red ladybug investigates a purple lupine (LOO pine) plant.

Monarch butterflies carefully lay their eggs on the leaves of milkweed plants. In a short time, caterpillars like this yellow-and-black one will hatch.

After a few weeks, each caterpillar surrounds itself with a hard shell and changes into a monarch butterfly. Soon it will fly from flower to flower, as gracefully and beautifully as any bird.

Flies are among the park's many scavengers, eating whatever they can find. Dead animals and leftover picnic foods are always among a fly's favorites.

The buzzing noise that a fly makes comes from the way it moves its wings. The wings of a common fly beat almost 200 times per second. But that's nothing compared to the wings of the small no-see-um flies that live deep in a forest. They move their wings more than 1,000 times a second!

Buzzing mosquitoes are plentiful. They make an easy feast for the blue jays that constantly dart from place to place.

Other birds work a bit harder for their food. This mother robin has finally managed to catch a worm for her youngsters. The high-pitched cheeps of the baby robins let their mother know just how hungry they are.

Worms are not the only creepy-crawly creatures in the park. One lucky snail has found a wild strawberry plant. The berries make an excellent meal.

The slithery garter snake, however, has not been so lucky. It must move on, searching for a mouse, a bug, a fish, or even an earthworm to eat.

Ants almost always find something to eat. They make their homes, called colonies, under the ground or between pieces of broken walkway. Ants are social creatures, which means they live and work in groups.

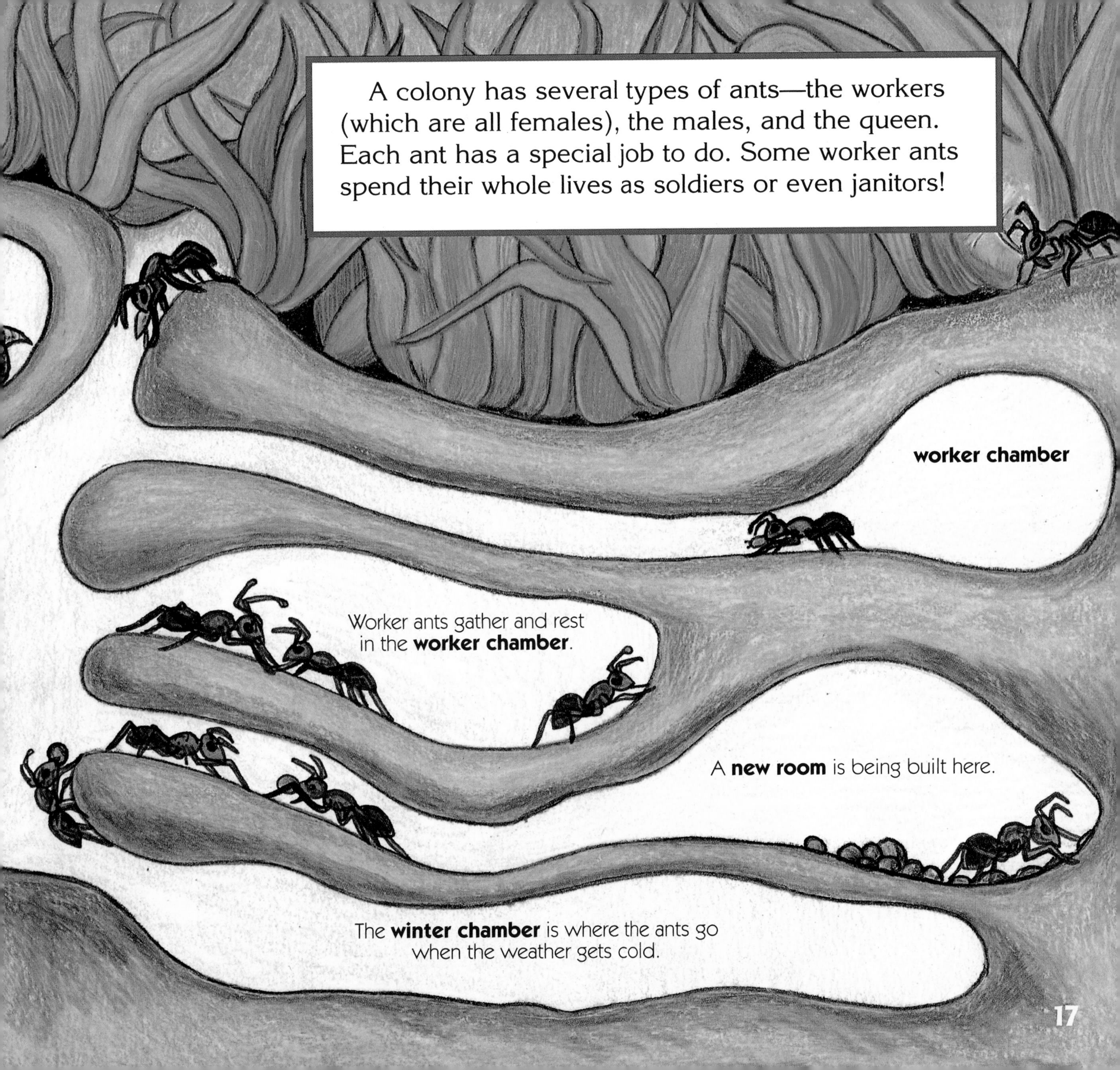
A colony has several types of ants—the workers (which are all females), the males, and the queen. Each ant has a special job to do. Some worker ants spend their whole lives as soldiers or even janitors!
worker chamber
Worker ants gather and rest in the **worker chamber**.
A **new room** is being built here.
The **winter chamber** is where the ants go when the weather gets cold.

Not far from the ant colony, beneath a large, gray rock, lives a round, squat toad. Usually he is a quiet fellow. But sometimes he breaks the silence with his high-pitched call, hoping to attract a mate.

On a nearby pond, a family of ducks swims and plays beside some tall, stringy reeds. Mallard ducks like these feed by dipping their heads in the water and swinging their tails up into the air. This is how they reach the most delicious underwater plants and animals.

Just beyond the pond, a mother swan watches over her youngsters with great care. Her mate is most likely close by. Unlike many other birds, swans mate for life, usually at age two or three. They take such excellent care of their offspring that young swans often remain with their parents until it is time to find mates of their own.

In the fall of each year, Canada geese visit the pond. They stop for a while, resting and feeding on nearby grass and plants. Then the geese fly off to spend the winter further south. But in spring they come again, heading back to their northern homes.

When they travel, Canada geese fly in a V-shape. This helps the flock of birds stay together. It also helps them fly long distances. As the birds flap their wings they push air up. All this moving air helps keep the birds aloft during their long flights.

These squirrels have been eating nuts, seeds, and wild mushrooms all day. As the sun sets, they take one last look around for food. Then it will be time for sleep.

Other park animals come out to eat and drink as well. Unlike squirrels, rabbits like this cottontail sleep most of the day. Then, from dusk to dawn, they play and hunt for food. Clover, grass, twigs, and tree bark are all delicious to a hungry wild rabbit.

Later, in the darkness of night, bats leave their roosts under bridges and streetlights. Winging their way through the trees, they spend the night hunting for insects of all kinds, especially mosquitoes.

Among the park's fiercest predators, or hunters, are owls. Their remarkable eyes allow them to find mice and other small mammals even at great distances. Owls also have very sensitive hearing. The rustle of a leaf is enough to tell an owl exactly where to pounce.

As the night goes on raccoons appear from their dens. They scavenge for food, hoping to discover good things to eat that someone has left behind. For a raccoon, a full garbage can is a wonderful treasure.

It is dawn of a new day, and the night creatures have gone back to sleep. Now the daytime birds, insects, and small mammals awaken. Among the early risers is this peregrine falcon. Down it swoops, diving from high on a skyscraper to the park far below.

Skyscrapers make excellent homes for peregrine falcons. There, these predators can have nests even higher up than ones they could make in trees. When they spy an unsuspecting pigeon or other small bird, the falcons dive down from their stone perches at speeds of up to 200 miles (322 kilometers) an hour!

While the streets come to life, pigeons fly to the park from streetlights, window sills, and telephone poles. Like many other visitors, they will find good things to eat on the paths and walkways.

The day passes, and visitors of all kinds come and go. Predators and prey, insects, birds, and people all manage to find something they need in the park. It's all there in this small green world hidden away from the cement and smoke and noise of the city.

More About

City Park, Page 7:
The world's largest city park is Fairmount Park in Philadelphia. It covers over 4,000 acres (1,600 hectares) and contains a zoo, several historic houses, and two theaters.

House Fly, Page 12:
To eat, a fly spits on its food. The spit dissolves the food so that the fly can soak up the meal with its spongelike mouthpiece.

Earthworm, Page 9:
An earthworm does not have eyes. Instead, it has light-sensitive cells in its skin. Most of these are in the front end of its body, which helps the worm find the soil surface.

Robin, Page 14:
A pair of robins usually raises eight babies during each breeding season. In 10 years, if all the birds survived, the offspring of this one pair of robins would total 19,500,000 birds!

Monarch Butterfly, Page 11:
Monarch butterfly caterpillars feed on the poisonous milkweed plant. Strangely, the poison does not hurt the caterpillars. It just makes them taste bad to predators.

Snail, Page 15:
A snail's eyes, which are on top of long stalks, can tell the difference between light and dark. This helps them find cover and know whether it is day or night.

This Habitat

Toad, page 18:
Toads lay jelly-coated eggs in the water. These will soon change into water-breathing tadpoles. After feeding on water plants, the tadpoles mature into air-breathing adult toads that eat insects.

Little Brown Bat, Page 24:
Bats may be tiny, but they have huge appetites. One little brown bat can eat 600 mosquitoes in a single hour.

Gray Squirrel, Page 22:
Squirrels depend on the park trees for food, shelter, and protection. In return, a forgotten acorn buried by a squirrel may grow into another oak tree.

Raccoon, Page 26:
A raccoon's front paws, which look like a child's hands, are useful for finding food in the park. These paws can lift garbage can lids, unlatch gates, and even untie string!

Cottontail Rabbit, Page 23:
Rabbit ears stand up tall and swivel to gather sounds from all directions. This lets rabbits listen for danger behind them while they nibble on grass.

Peregrine Falcon, Page 27:
Peregrine falcons were once threatened by the pesticide DDT. Now DDT has been banned, and people are helping falcons through breeding and release programs.